The Sounds We Hear

by Caroline Hutchinson

Sounds are all around us.
People can hear some sounds.

What are some sounds that people can hear?

People hear sounds in many places.

People hear sounds in the city. People walking and talking make sounds.

Car doors make sounds when they open and close. Horns and sirens make sounds, too.

What sounds can you hear in the city?

People can hear many sounds in the country. Sounds can be loud. A tractor makes a loud sound.

Sounds can be soft, too. A cat makes a soft sound when it is happy.

A sound can be loud or soft.

People can hear sounds at the beach. The water makes a loud, low sound when it washes up on the sand.

Some sounds are high. This bird makes a loud, high sound.

A sound can be high or low.

Sound is energy that moves as waves. Things make waves when they move back and forth.

The faster something moves back and forth, the higher the sound.

Sounds are vibrations.

You can't see the waves as they move, but some computers show how the waves could look.

These waves show sounds.

Sounds move as waves through air. Sounds can move through water and the ground, too.

Put your fingers on your throat and talk. Can you feel the parts in your throat move back and forth?

You can't see the waves when a lion roars, but you can hear the sound!

Lions can make sounds.

People hear many sounds.
People hear sounds in many places.